Design Considerations for a Software Space Elevator Simulator

International Space Elevator Consortium
Autumn 2017

Authors:
Dennis H. Wright
Steven Avery
John Knapman
Martin Lades
Paul Roubekas
Peter A. Swan

Design Considerations for a Software
Space Elevator Simulator

Published by Lulu.com

dennis.wright@isec.org

978-1-387-65437-6

Printed in the United States of America

Preface

The vision of the International Space Elevator Consortium (ISEC) is to have

> "a world with inexpensive, safe, routine, and efficient
> access to space for the benefit of all mankind."

As a necessary step towards achieving this vision ISEC has undertaken a series of year-long studies, each of which focuses on a particular aspect of the design and construction of an operational Earth-based space elevator. The 2017 study deals with the requirements and preliminary design aspects of a software simulator of a space elevator system. The goals of the study are to identify the most important functions of such a simulator, derive from these the requirements of the software to be developed and outline its major design characteristics. The resulting study report will serve as a guideline for development, rather than a detailed blueprint, so that the software can evolve as needs arise. The end goal is a "gold standard" simulation toolkit to be used throughout the space elevator community.

The authors of this report wish to thank the members of ISEC for their support, Robert E. 'Skip' Penny, Jr. and Peter Robinson for their contributions to this report, and the attendees of the 2015, 2016 and 2017 ISEC Conferences for ideas contributed during discussions.

Signed: *Dennis H. Wright*
             ISEC Director of Studies, Editor
             31 January 2018

Executive Summary

As with all large, modern engineering projects, detailed computer simulations of the space elevator will be essential during its design, construction and operational phases.

Within the context of these phases, this study enumerated 14 use cases which the simulation software must address, ranging from 3D dynamics and electrodynamics calculations of space elevator motion, to the effects of payload capture and release at various points along the tether, to the effects of friction arising from the interaction of the space elevator climber with the tether. Proceeding from these use cases, requirements were imposed on the software design and an outline for its development was sketched.

A central part of the design is a general math and physics platform which can perform the many calculations required. The study team reviewed seven such platforms and chose Mathematica as the one most likely to meet the needs of the simulation. To maintain an open-source option, SageMath was chosen as an alternative math/physics platform. Applications specific to the space elevator simulation will be built on top of these platforms.

The simulation software must be developed using modern, best programming practices, and employing Model-View-Controller (MVC) design so that all but a few of the many details of particular space elevator applications are hidden from the user. The simulation must also be modular and flexible enough to evolve with the changing needs of its users.

Finally, the software must be made available to a variety of users through various distributed computing technologies such as the cloud. Security issues must be addressed throughout the design and implementation of the software and maintenance will require periodic upgrades and regular testing.

Based on these findings, the study team made 11 recommendations concerning the space elevator simulator. The major ones are:

- A software space elevator simulator should be developed.
- It should be based on Mathematica and SageMath.
- It should be professionally developed and maintained.
- Its development should be funded by a crowd-funding campaign.

Table of Contents

1 Introduction

A standard part of a modern engineering project is the computer simulation of its major components, both individually and as a co-working whole. This is especially true for large or complex projects in which physical prototyping is very expensive, time-consuming or, in some cases, impossible. Considering that the space elevator is both very large and very complex, extensive computer simulation of its behavior during prototyping, deployment and operational phases is essential.

Recommendation 1: A software toolkit should be developed which can simulate the space elevator.

The International Space Elevator Consortium (ISEC) (see Appendix A) therefore initiated a study to specify the requirements and design aspects of a software toolkit which could simulate all major features of a space elevator. The goals of the study were to:

- Identify the most important functions of the simulator.
- Derive from these the requirements of the software to be developed.
- Outline the major design characteristics of the software.

The report from this study is intended to serve as a guideline and not a detailed blueprint for the future development of a software simulator. It was clearly understood that any such software will evolve as understanding increases and new needs arise. It is a goal of ISEC that the software resulting from this study should provide the high-quality tools needed for the development of new space elevator applications and a standard means of comparing new and existing space elevator models with data and with each other.

This document is the eighth in a series of yearly technical reports, each of which deals with a specific aspect of space elevator development or operation. A list of these is presented in Appendix B. They are available either for sale in hardcopy or free as pdf files at www.isec.org. Each report was produced using the ISEC study process.

1.1 The ISEC Study Process

ISEC developed a process of selecting a key topic for in-depth analysis and then conducting a year-long study to assess various aspects of the topic. This enables ISEC to prioritize activities and leverage the expertise of volunteers in the relevant fields. The focus on a single topic for a particular year enables the community to bring together its strengths and address the topic at the yearly conference. The process culminates with a report which makes recommendations for future action.

1.2 Report Layout

Chapter 1, Introduction, discusses the necessity of a space elevator simulator and sets forth the goals and process used to guide the study and generate this report.

Chapter 2, Context, Scope and Use Cases, discusses the situations in which the space elevator simulator will be used, and what aspects will, and will not, be simulated. This chapter also considers specific questions which the simulator will need to answer.

Chapter 3, Requirements, describes the features that the simulation software must have and how they follow from the listed use cases. Requirements are categorized as either functional, which directly affect simulation outcomes, or non-functional (e.g. open source), which do not directly affect outcomes.

Chapter 4, Design, lists guidelines for software design and outlines the overall architecture which follows from the requirements. Brief discussions of the software architectural concepts of object-oriented design and model-view-controller are included.

Chapter 5, Implementation, outlines the proposed software development, testing and support. A concept of operations, covering methods of collaborative program development and use, as well as software distribution and maintenance, is also discussed.

Chapter 6, Roadmap for Development, makes suggestions for how simulator development should be sequenced and funded. A possible development plan is laid out with a very approximate cost estimate. The evolution of the software over time and its administration is also discussed.

Chapter 7, Conclusions and Recommendations, summarizes the main findings of this report and makes recommendations to ISEC for how the simulation software should be developed.

Appendices briefly describe ISEC, define several terms and acronyms, list past ISEC studies and provide minutes of the brainstorming sessions dedicated to this topic.

2 Context, Scope and Use Cases

The space elevator simulator will be used as a software tool in the research, design and operational phases of a space elevator system. Each of these phases defines a context in which questions can be posed and answered. Enumerating which questions can and cannot be answered within a given context defines the scope of the software and avoids tangential development. Given context and scope, specific use cases can be defined.

2.1 Context of a Space Elevator Simulator

Four main arenas of study define the contexts in which the simulator will operate: space elevator dynamics, interaction of the space elevator with its environment, interaction of tether and climber and normal operation and failure modes

An overview of the dynamical system that must be simulated is shown in Figure 1. It includes tether and climbers, the Earth Port, GEO node and Apex Anchor. For definitions of these and other terms please see Appendix C. Shown here are a number of effects that must be included in the simulation in order to obtain good predictions of space elevator motion: Earth Port and Apex Anchor masses and motions, tension and elasticity in the tether, gravitational stabilization, motion and masses of tether climbers and wind forces in the atmosphere.

Figure 2 places the space elevator within the context of its environment. The length of the tether is shown approximately to scale with respect to the Earth and its magnetosphere. Also shown is the solar wind and bow shock into which the space elevator and Apex Anchor extend during part of the daily rotation. Half a day later the entire space elevator is enveloped within the magnetosphere. Thus the electromagnetic and radiation fields through with the space elevator moves are continuously changing, giving rise to highly dynamic effects which must be simulated. A precise model of the Earth's gravitational field is of course also required as are lunar, solar and, to a lesser extent, planetary fields.

The interaction of a climber with the tether is a complex topic whose simulation requires the coupling of several diverse areas of physics. A representation of a possible tether gripping mechanism is shown in Figure 3. The friction between the climber gripping mechanism and the tether is critical to the operation of the space elevator. Several things about this process are unknown and must be studied: the coefficient of friction of the tether material, heat transfer from the climber to the tether, the effect of compression and expansion of the tether as the climber passes by, and so on. Repeated gripping by the climber will likely degrade the tether material as will the impact of small particles of space debris.

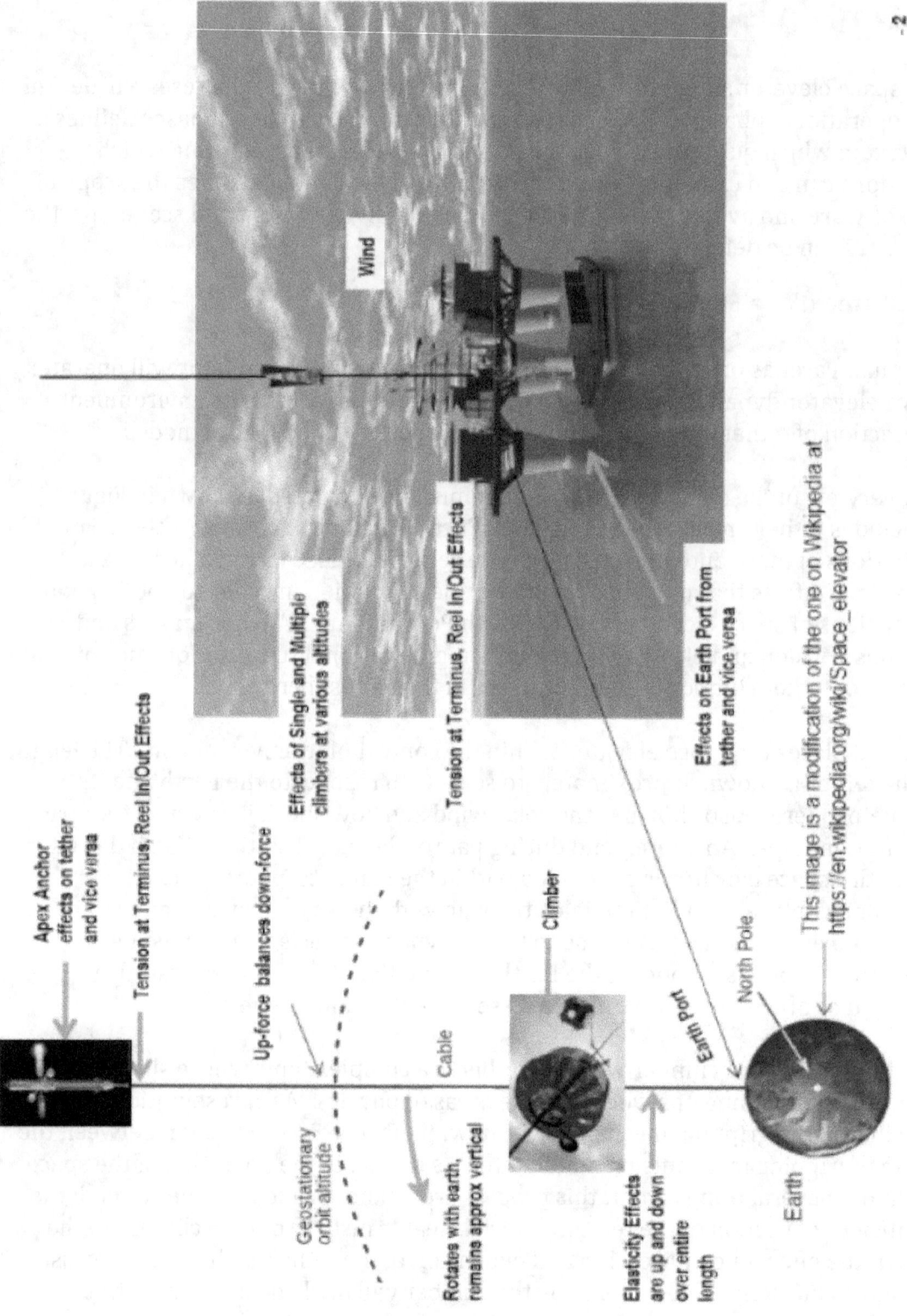

Figure 1: Overview of the Earth-based elevator dynamical system to be simulated under normal operational conditions.

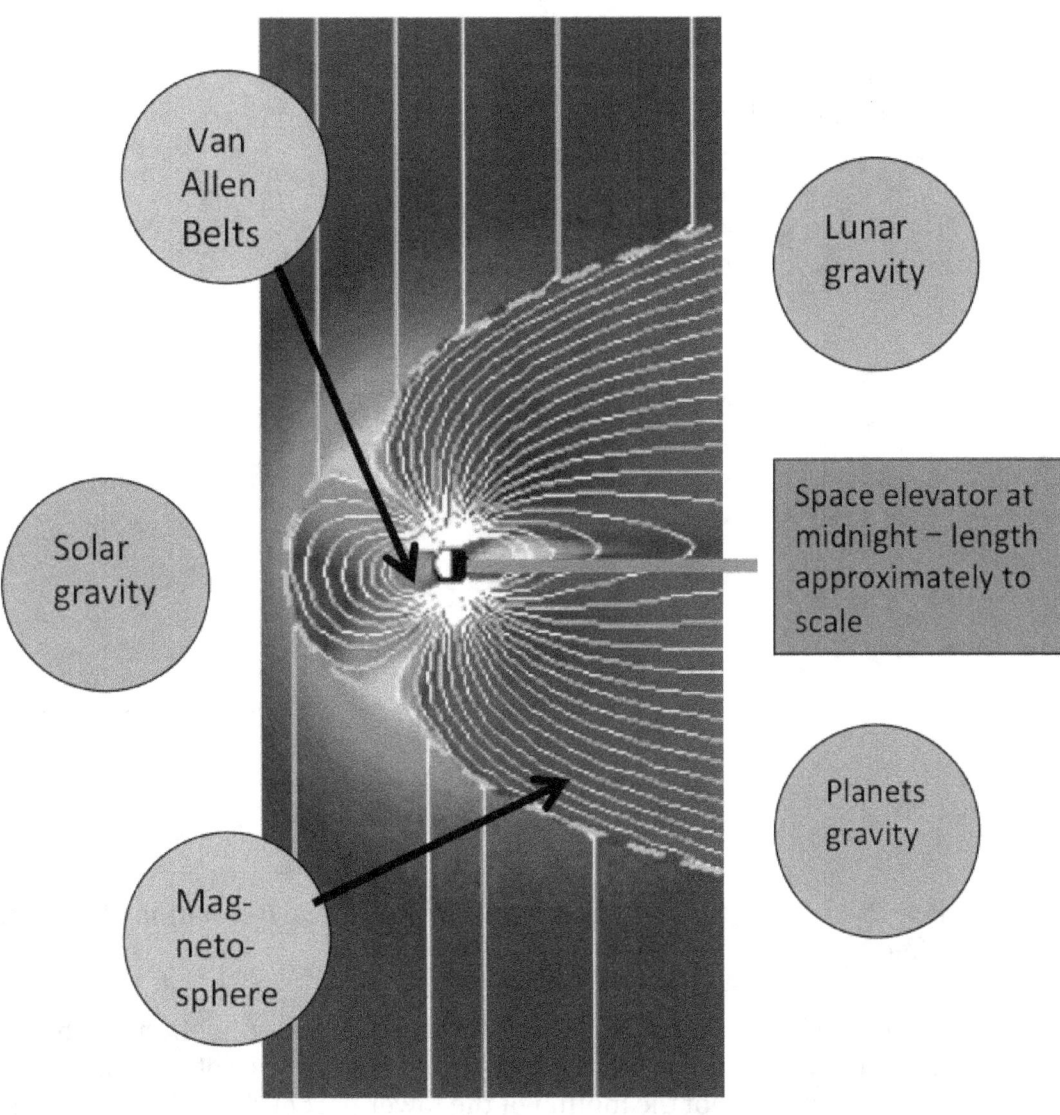

Figure 2: The electromagnetic, gravitational and radiation environment in which the Earth-based space elevator will operate.

The simulator will play a major role as part of a feedback control loop. During construction and normal operation conditions tether and climber positions, tether tension and other parameters will be monitored and fed into an elevator control system, which will include a simulation model. This complex system will use the model to determine the optimum control outputs to yield the desired elevator system motion.

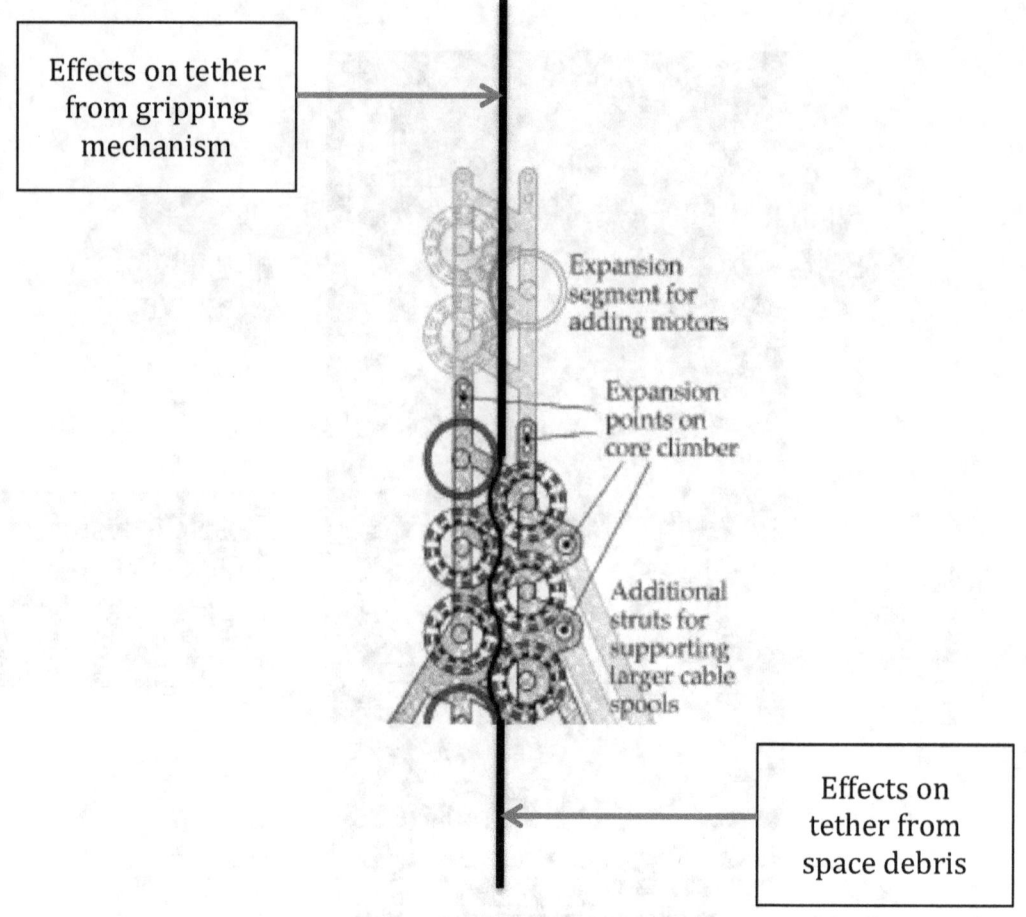

Figure 3. Possible tether gripping mechanism showing the deformation of the tether and other effects.

Operations will also involve planning for failure modes. The simulation will need to predict unexpected motions or events, one of which is severance of the tether. Figure 4 is a representation of the motion of the lower part of the tether after it has been severed, perhaps by a larger piece of space debris. Understanding what happens to all parts of the space elevator in such an event will inform recovery and repair procedures, and help to forecast damage caused by uncontrolled pieces of the tether.

Recommendation 2: The simulator should serve and inform the development, construction and operational phases of the space elevator.

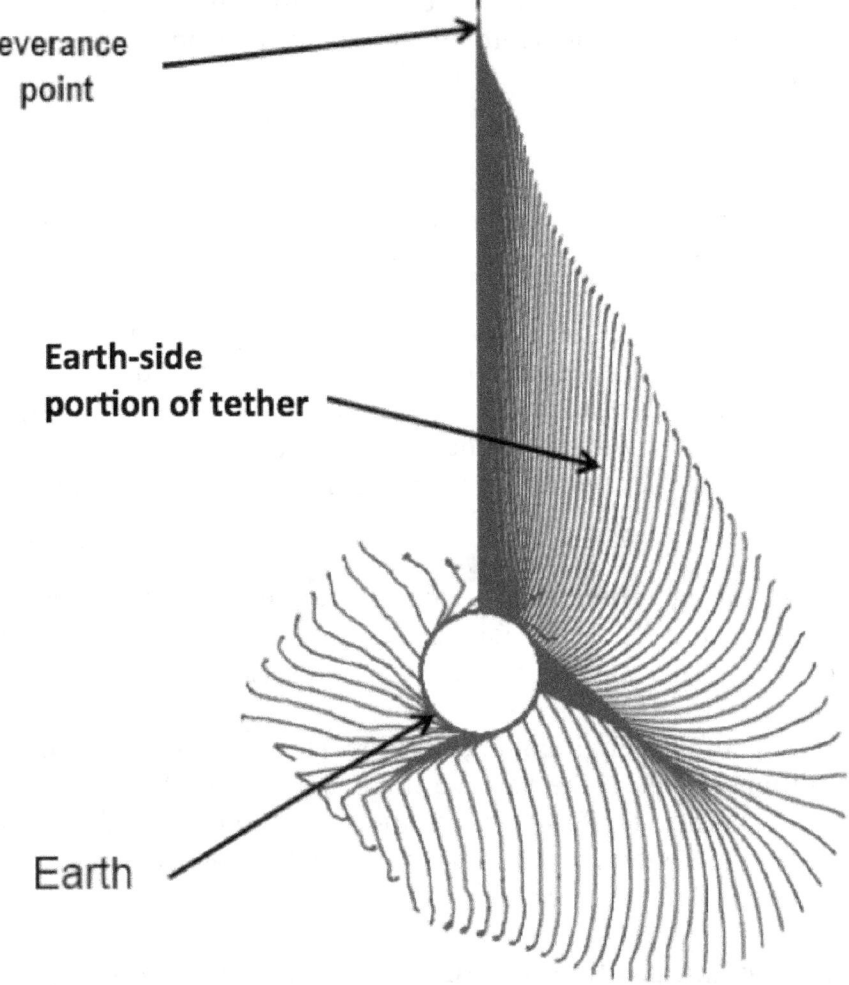

Figure 4. An example of failure mode analysis: the time evolution of the lower part of the space elevator tether after a severance. Calculation by Paul Williams.

2.2 Software Scope

Many different simulation models will be required to support the construction and operation of a space elevator, but not all will be included in the current project. A simulation model will not be included if it either has no direct influence on space elevator motion, or its influence is sufficiently indirect that it can be factored into another, distinct software application. Examples of such models are the distribution of space debris and its likelihood of impacting the space elevator tether, the traffic to and from the Earth Port, GEO Node and Apex Anchor, weather conditions at the Earth Port and in the upper atmosphere, Earth surface motions due to wave action, and economic and political forecasts.

While all of the above models could inter-operate with some future dynamics simulation model, direct inclusion into the project would lead to unnecessary complication at this stage.

2.3 Use Cases

Some general and specific use cases are presented here in order to survey the needs of a space elevator simulation software project.

Predict motion transverse to tether axis. Assuming the space elevator is in its operational mode, predict the tether's oscillation modes and side-to-side motions in response to stretching, climber motion, gravitational forces, electromagnetic forces or debris impact. This will be needed for studies of payload release, vehicle rendezvous, space debris avoidance, Earth port, GEO station and apex anchor positioning.

Predict motion along tether axis. Calculate tether stretching and contraction about equilibrium, and its oscillation modes, given imparted energy and tether material properties. This applies to the same studies as the previous use case.

Estimate effects of all engineering strains (tension, torsion, bending and shear) on tether. What is the relative importance of each strain, how do they couple with one another to affect tether motion, and what limits do they impose on tether operation? A model which handles this would likely also be general enough to accommodate the first two use cases.

Calculate tension and position of tether at Earth port. How will the Earth port be moved and what tension must it counteract due to vibrations and extensions/ contractions of the tether? Conversely, how will moving the Earth port affect tether motion?

Simulate reel-in/reel-out of tether at Earth Port and/or Apex Anchor. How does gathering in or paying out lengths of tether affect the motion and tension in the tether? Is it necessary to do this also at GEO or the Apex Anchor?

Simulate payload capture and release at various points along tether. Extra masses representing payloads, being attached to and detached from the tether, will cause local deflections from tether vertical and local variations in tether tension which will be propagated up and down the tether. These will need to be calculated and understood.

Calculate the effect of friction between the climber and tether on the bulk material of the tether. The climber will grip the tether and transmit heat energy into the tether. How will this affect the structural integrity of the tether?

Calculate effect of moving climbers on tether. What motions are induced when several climbers are in motion both up and down the tether, and what are the limits on climber speed?

Simulate severance of the tether. What happens to the upper and lower segments of the tether if it is parted at some point? The detailed motion of each piece along its length should be calculated: this data could then be used for risk analysis and optimization of sever mitigation actions.

Estimate effect on tether of its regular motion through the magnetosphere and solar wind. Electromagnetic forces and currents will be induced according to the electrical properties of the tether material. How is tether motion affected? This would include the effects of solar storms, for example.

Simulate the effect of space radiation on tether motion, bulk material, and climbers. How do fluxes of energetic charged particles affect the motion, surface currents and structural integrity of the tether? This would include light pressure from the scattering of photons, collected charge due to passage through radiation belts and induced radioactivity.

Calculate the effect of winds and other atmospheric effects on tether motion. Atmospheric winds can have a large effect on tether and climber motion and integrity of the tether.

Calculate tether motion when coupled to High Stage One. One end of the tether may connect to a high-altitude platform supported by various means. The platform and its support may take many different forms requiring different simulation approaches, but the mutual motion of tether and platform could be simulated with models already mentioned above.

Simulate the effect of lunar, solar and planetary gravity on tether motion. Any elevator tether will be subject to multiple gravity forces (lunar, solar, etc.), requiring the inclusion of, or access to a solar system ephemeris database in the model. This will require a careful choice of coordinate reference frame to ensure that all tidal and resonance forces are correctly modeled. Effects of a non-spherical Earth should also be included.

3 Requirements

In software design terminology there are two types of requirements: functional, which directly affect outcomes, and non-functional, which do not. The in-scope use cases discussed above drive the functional requirements. Non-functional requirements deal with the way the software will operate or be maintained or be distributed, among other things. A non-exhaustive list for each class of requirement is presented here.

3.1 Functional Requirements

Dynamics simulation of all engineering strains: includes tension, torsion, bending and shear and all couplings between them.

Electrodynamics simulation: includes the forces exerted by electromagnetic fields on the space elevator tether and the currents and voltages induced.

Simulation of radiation effects: the effects of solar radiation, cosmic rays and radiation belts on the tether material and motion.

Faithful representation of the physics environment: includes gravitational fields, electric and magnetic fields, radiation fields, atmospheric effects and so on.

Capability to model friction: both static and rolling.

Direct access to databases: such as gravitational, radiation distributions, etc.

Validation against real-world data: includes comparison to space-based and Earth-based experiments.

Software must be versatile. It should accommodate different space elevator models, some of which already exist.

Pipelining of results from one model to another should be possible. This would allow run-time cross-comparison and validation.

3.2 Non-functional Requirements

Simulation software may be proprietary or open source.

A standard schema should exist for comparing results of different models. This makes objective evaluation of results easier.

Users must have access to model parameters. These parameters may represent a restricted set of the full parameter space.

Longitudinal testing must be done. Routine testing of the software over time helps to understand changes and provide quality assurance.

Software verification must be performed. Tests should be developed which demonstrate that the software is doing what it claims to do. In the long term this verification must include comparison to real-world test and operational data.

High-quality visualization should be available. This includes real-time displays of predicted space elevator motion, monitors of various functions as model parameters vary, and movies.

Analysis tools must be available. These are needed to histogram or plot results and to extract results from simulation applications.

Non-functional requirements pertaining to the development, operation and maintenance of the software are discussed in more detail in Chapter 5, **Implementation**.

4 Design

This report does not intend to lay out a detailed design of the space elevator simulation software; instead it will outline the major design elements required for flexible, evolving software. At this writing it is not known what portions of the software will be developed by professionals or by talented volunteers, or both. Nevertheless, adherence to a few principles will help to ensure high quality.

4.1 Design Guidelines

Three modern programming concepts will be followed. These are object-oriented (OO) programming, model-view-controller (MVC) design and Unified Modeling Language (UML) diagrams.

4.1.1 Object-oriented programming

Object-oriented programming is based on "objects" which contain data and instructions which operate on the data. Software projects are then built from a collection of objects which can interact with one another in well-defined ways. Programming in this manner encourages better association of data with functionality, better segregation of unrelated parts of the software and more modular design.

4.1.2 UML Diagrams

UML diagrams are a useful tool in OO design. They provide a standard way to visualize the structure and relationships of objects:

- Objects are represented by boxes which contain lists of data held in the object and operations that the object can perform.
- Boxes are connected by various types of line which indicate relationships between objects.
- Groups of related objects may be collected into and represented by package (or file folder) shapes.

A basic set of UML diagrams follows from the requirements and use cases and allows major design decisions to be made before any software is written. Once the design is settled, specific programming rules (such as those in C++) allow a direct translation of UML diagrams into software.

Specific UML diagrams for the simulator project are shown in section 4.2.3.2.

4.1.3 Model-View-Controller Design

MVC is an approach to software architecture which divides application software into three components: models, which describe solutions to specific problems, views, which send data to the user based on changes in the model, and controllers, which take input from the user and send it as commands to the models.

For space elevator applications, a model could be a representation of tether motion as a function of time, a description of the Earth's magnetic field or a dynamic formulation of energy imparted to the tether by the gripping of the climber.

Views describe user interfaces, graphical and otherwise, which present model output to the user so that it can be analyzed. Analysis could take place visually, through the use of histogramming or plotting tools, or by numerical methods.

Controllers allow the execution of models and views to be steered through the use of a restricted set of parameters, while hiding implementation details that are likely of no interest to the user. Controllers together with views provide the user interface, which may take the form of a control console or command line reader.

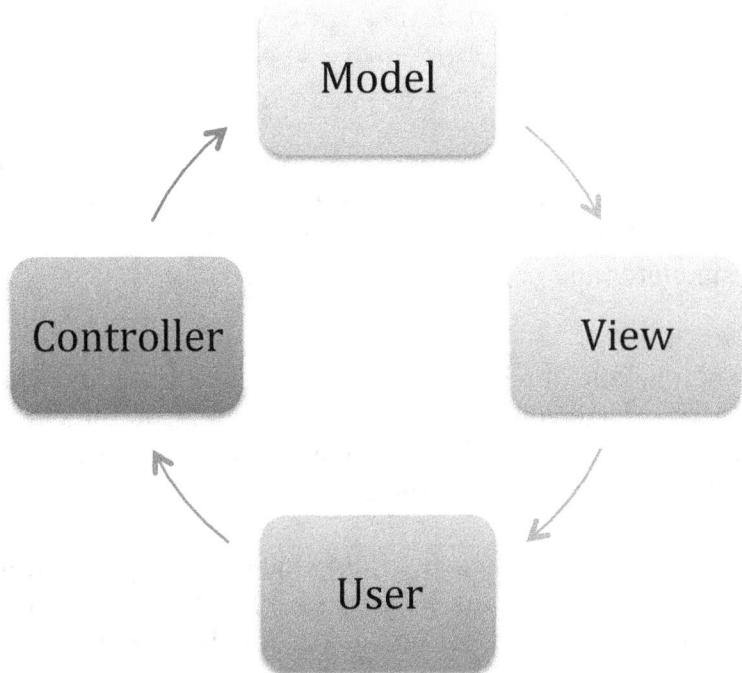

Figure 5: Model-View-Controller concept. Controllers interpret input from the user, models respond to commands from controllers and views send output from models to the user.

As shown in Figure 5, this type of architecture separates the details of the models from the user who typically will only wish to change a small number of model parameters and observe the result.

Recommendation 3: The design of the simulation software should follow the Model-View-Controller (MVC) concept.

4.2 Software Architecture

From the requirements discussed in Chapter 3, it is clear that a wide variety of physics and mathematics algorithms must be included in the simulation. Because such algorithms are usually CPU-intensive, they must be efficient and optimized for the tasks at hand. Rather than re-invent the wheel, it makes sense to take advantage of math/physics platforms which already supply most of the needed functionality. Building a simulator on top of one or more of these platforms would save years of work, ensure a degree of standardization and enhance reliability.

The math/physics platform will need to work seamlessly with space elevator applications, databases and user interfaces. The linkage that makes this possible will be supplied by a set of software modules called a framework.

Recommendation 4: A multi-purpose math/physics platform should be chosen upon which the simulation software will be built.

4.2.1 The Math/Physics Platform

The study team considered several general platforms now used in science and engineering, as well as some specific to space and tether applications. These platforms were evaluated according to criteria deemed important for a general space elevator simulator and one was selected.

4.2.1.2 Selection Criteria

The calculation of full, four-dimensional motion must be available. This includes longitudinal and transverse vibrations and their damping, non-linear motions near to and far from stability, motions of moving masses (climbers) coupled to the tether and the effects of climber capture and release.

Physics capability must be extensive, providing the modeling of tension, torsion, shear and bending, Young's modulus, Poisson's constant, coefficient of friction and other mechanical properties of the tether, and the modeling of gravitational, electromagnetic and radiation effects, conductivity and other electrical properties of the tether.

Mathematics capability must include non-linear, highly coupled differential equations, fast numerical integrators and numerous advanced math functions. It must also be able to perform finite element calculations.

Modules to model environmental effects such as atmosphere, tides, earthquakes and ocean waves should be available.

The platform should be compatible with external software such as libraries of other physics software, external databases and existing space elevator models.

Benchmarking, regression testing and unit testing should all be supported.

Cost should be reasonable and licensing should accommodate world-wide usage of the software without US export control (ITAR) restrictions. Open source software is preferred but not required.

Excellent support of users and software developers is desired.

The platform software should be easy to build, use and maintain.

Direct experience with the platform by ISEC members is very important.

4.2.1.3 Survey of Existing Platforms

Seven multi-purpose math/physics platforms were considered. Two, GTOSS and TetherSim, are used in space and tether applications. Five are for general science and engineering use: MATLAB, ANSYS, COMSOL, SageMath and Mathematica. Each platform was measured against the above criteria where possible.

GTOSS is a well-known heritage application for simulating complex tether-and-mass combinations and has been used for many NASA applications. It is free of charge and open-source. It supports tether calculations in gravitational and magnetic fields, but does not support continuum mechanics calculations, bending or torsion strains. The user interface consists of a simple input file and periodic data dumps during the execution of the program. Rather than a general simulation platform, GTOSS would serve more as a specific type of space elevator application.

TetherSim is a modern simulation tool for tether dynamics which is currently being used for several space applications. It is more general than GTOSS and supports continuum mechanics, electrodynamics and all engineering strains except torsion. It is proprietary software costing several thousand dollars per license, including user support. It is not clear if this product is easily compatible with external applications or with space elevators significantly different from the kind it is designed to deal with.

MATLAB is a programming language especially designed for general numerical calculations. Some dedicated physics modules are available but none are especially useful for space elevator physics; users must program what is needed. It supports object-oriented programming, is somewhat similar to C++ and is compatible with programs written in other languages. The base software is not open, with licenses costing around $2000 per user. MATLAB has been used for 2D space elevator simulations, but 3D cases are laborious and difficult to implement. It has no user interface and is not seen as user-friendly.

ANSYS is a professional simulation tool widely used in industry, including aerospace and space systems. A comprehensive set of specialized physics modules provides the structural mechanics, electrodynamics, differential equation solvers and 3D finite element analysis needed to simulate the space elevator. ANSYS is flexible, compatible with external software and relatively easy to use after sufficient training. Good user support is available. The software is proprietary, costing up to $30,000 per user license and varying from country to country, but quotes are made privately to individual customers.

COMSOL is a general modeling platform used to simulate electrical, mechanical, fluid flow and chemical applications. Its multi-physics approach allows various physical effects to be treated in a coupled way. Some add-on modules, which cost extra, would be required for space elevator simulation. There is no gravitational module so it would have to be developed by users. Users can insert non-linear differential equations for custom modeling and finite element analysis is included. COMSOL has a good user interface, but more complex physics modeling is not particularly user-friendly. It is not clear if COMSOL libraries are compatible with those of other modeling toolkits. Its cost is $30,000 per user license, which is valid world-wide and in perpetuity. Good user support is supplied.

SageMath is free software for mathematical modeling and is intended as an open-source alternative to MATLAB or Mathematica. It is not designed in terms of specific physics modules, but rather as a more general mathematics toolkit providing differential equation and linear algebra solvers. It currently does not support finite element analysis. Several graphics modules are supplied for plotting and analysis. The coupling of physics tools is not provided, requiring users to develop this feature. The SageMath libraries can be used interchangeably with those of Mathematica. Although there is no professional user support, a community of SageMath developers answers questions on a best-effort basis.

Mathematica is a general programming platform for mathematics, engineering and other disciplines. It contains a suite of math solvers for differential equations and finite element analysis, among others. It handles partial, non-linear and high-order differential equations in a general way. Like SageMath, it is not designed around specific modules, but rather as a general math toolkit. It does, however, contain an aerospace module and some features for dynamic and electrical computations. Users are generally responsible for developing applications for their specific use cases. Mathematica libraries are very compatible with those of SageMath. Mathematica is relatively inexpensive, with single user licenses costing about $300 and institutional licenses in the several thousands. Professional user support is included.

4.2.1.4 Recommended Math/Physics Platform
The study team decided that a platform should be general and not focused on a particular discipline or set of applications. It was therefore decided to remove GTOSS and TetherSim from the list of candidates.

While ANSYS and COMSOL are certainly general, they seem to be aimed mainly at institutional engineers and may not be particularly amenable to research applications with fairly deep math requirements. Their high cost would be a barrier to individual users who wish to contribute to space elevator development.

This leaves MATLAB, SageMath and Mathematica. MATLAB's lack of a user interface and the reported difficulty in setting up 3D simulations are strong negatives. SageMath seems much better, but it does not support finite element solutions for space elevator motion. However, its generality and library compatibility with Mathematica make SageMath an excellent open-source alternative which could be offered along side another modeling toolkit.

We therefore selected Mathematica as the math/physics platform because of its generality, low cost and compatibility with an open-source alternative. SageMath will also be made available in order to provide this alternative. In order to handle cases where Mathematica provides functionality that SageMath does not, standards for data interchange between the two platforms should be developed. Work elements could then be performed in SageMath, with their results fed into Mathematica where additional work would be done.

Recommendation 5: Mathematica should be used as the math/physics platform of the simulator, with SageMath as an open-source alternative.

4.2.2 The framework

The framework can be viewed as the software analog of an electrical bus to which components may be attached and through which they can communicate with one another. It is responsible for organizing the many software modules required for simulation and making them work together. General purpose frameworks such as USQUE already exist, but the choice of framework, and whether or not a custom version is required, will depend on the math/physics platform chosen.

4.2.3 Design Outline

4.2.3.1 Software to Be Written

None of the software in our survey supplied all the features that will be required for a versatile space elevator simulator. Many features will need to be added in order to supply all the necessary physics, on-demand database access, flexible user interfaces and access to already existing space elevator models. Applications addressing the use cases listed in Chapter 2, and those not yet anticipated, will be written using the functionality provided by Mathematica and SageMath.

A number of service, administrative and security functions will also need to be developed so that the entire simulator software can easily be used, distributed and maintained in a modern and evolving computing environment.

The first step in the design process will be to develop an architecture for the simulator software in which all of its major components and their relationships to one another are laid out and detailed. In a good design, the major components will be modular enough that they can be developed independently of one another. Each major component should in turn be laid out and detailed before software development can begin.

4.2.3.2 Preliminary UML Design

A set of UML diagrams was developed based on the above guidelines and the recommendation to use existing software to provide general math and physics functionality for application development.

Figure 6 shows the package-level design. It indicates the largest, highest-level components and how they are related, with MVC design playing the central role. The interfaces to Mathematica and SageMath provide access to the basic math and physics functionality required by the models. Physics databases refer to interfaces which access publicly available, professionally maintained collections of data on gravitational, electromagnetic and radiation fields which can impact space elevator motion. The resources package deals with CPU issues, memory allocation and storage, while the utilities package is as yet undefined. The system package could eventually depend on both of these. View elements refer to all the components necessary to develop views and display results to the user.

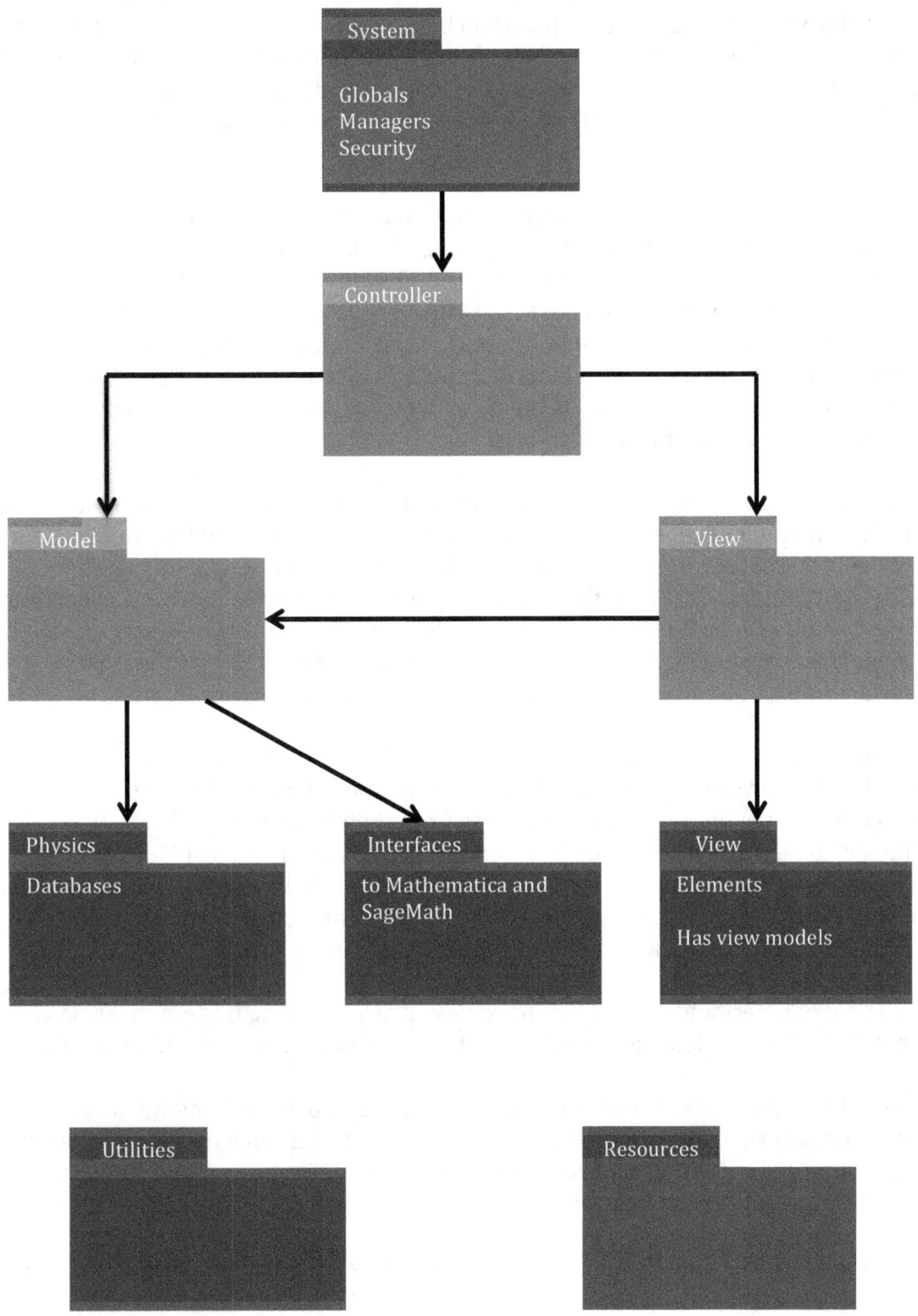

Figure 6. UML diagram of simulator packages showing dependencies. An arrow pointing from package A to package B indicates that A depends on B.

It is useful to look in detail at a few of the packages in the above design, specifically the model, view and controller. These will be described using some of the terms and elements of the C++ programming language. In the following UML diagrams, boxes represent *classes* of *objects*. A class is a C++ construct defining data and operations on the data. An object is an instance of a class which corresponds to a specific location in memory.

Classes may be grouped in well-defined relationships using a feature called *inheritance* in which common features of several classes may be abstracted out and placed into a higher-level or ancestor class. Common programming then need not be repeated in the descendant classes, and in some cases related descendant classes may be used interchangeably. Inheritance is designated in UML diagrams by an arrow pointing from the descendant or *derived* class to the ancestor or *base* class. Listed at the top of the class box is the class name and below that, the operations that the classes can perform, called *methods*.

Model Package The model package will contain all the software representations of physical components of the space elevator tether, climber and environments. There will likely be several different implementations of each component, of varying degrees of complexity. The tether, for example, may be represented by a collection of rigid rods, a single flexible ribbon or a series of springs. All of these are related by common properties and could be used interchangeably by sharing the same interface. This relationship is expressed in the class diagram in Figure 7.

The Model class is the most general and highest-level class from which all other model classes are derived. Models become more specialized as the diagram is descended, but maintain common features from their base classes. So ClimberModel, TetherModel and FieldModel classes are all kinds of Models, and RigidRods and ContinuousRibbon classes are kinds of TetherModel. This scheme of inheritance accommodates a wide range of different models while preserving commonality and re-using common program segments.

The chain of inheritance can be extended downward until the bottom-most class represents the fully detailed implementation of a specific climber, tether or field.

External models, that is, those developed by other authors not using the tools of this simulator, will also belong to the model package. For each such model an interface will need to be written which allows the external model to be used as if it were native to the proposed simulator.

View Package The view package contains the classes required to take output from the models and display it to the user. Many different views should be available, including complete or partial views of the space elevator and its motion, data feeds from the model to the user's monitor, and graphs and histograms updated as the model calculations proceed.

One kind of view might be a monitor which displays various aspects of space elevator motion. It may show climber motion or perhaps a large-scale view of the space elevator in a dynamic magnetosphere.

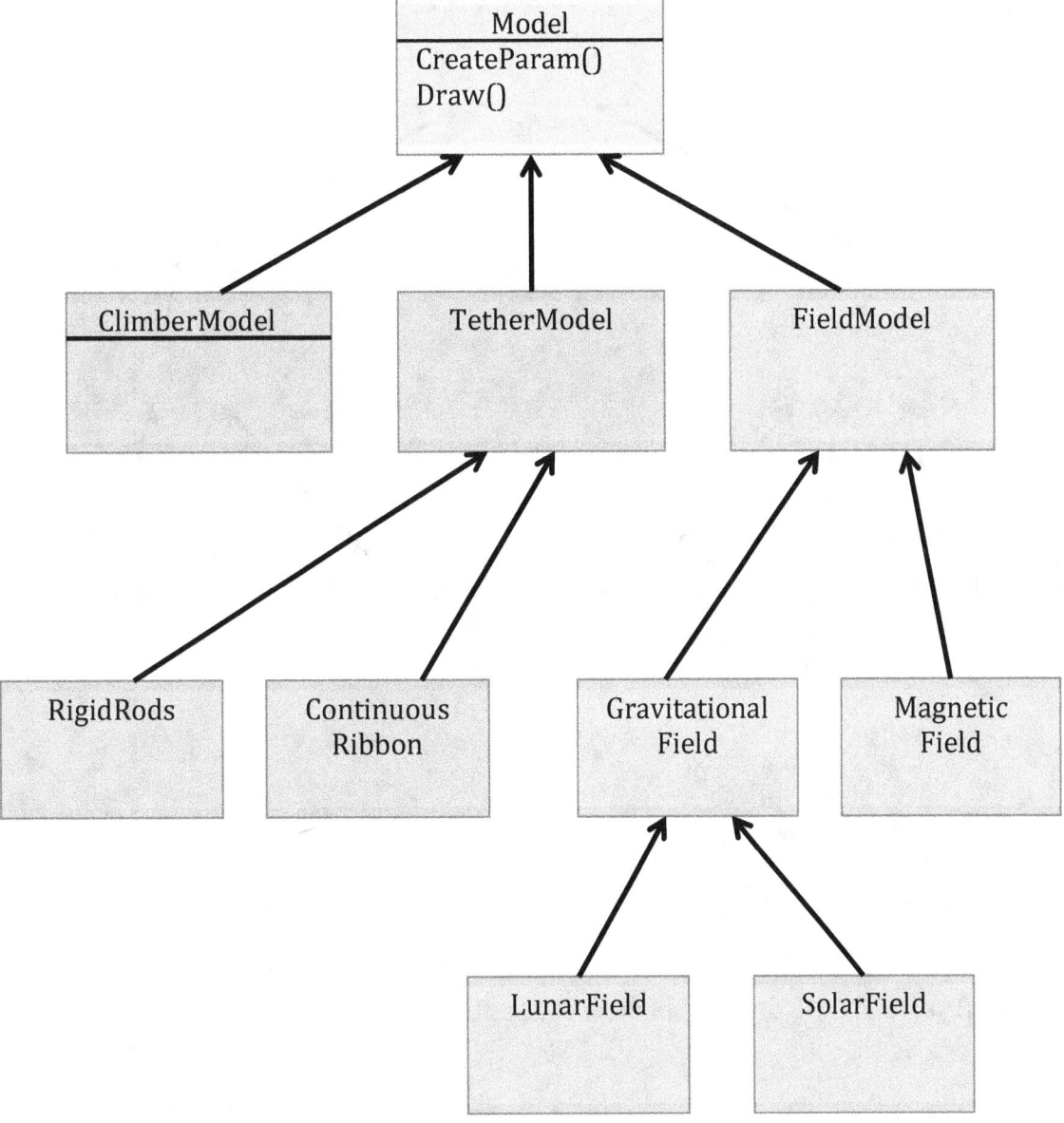

Figure 7. Partial UML diagram for model classes.

Views could also be used to monitor data flow through the use of dials and gauges, as well as the quality of the data feed.

Some of the classes needed to implement views are shown in the UML diagram in Figure 8.

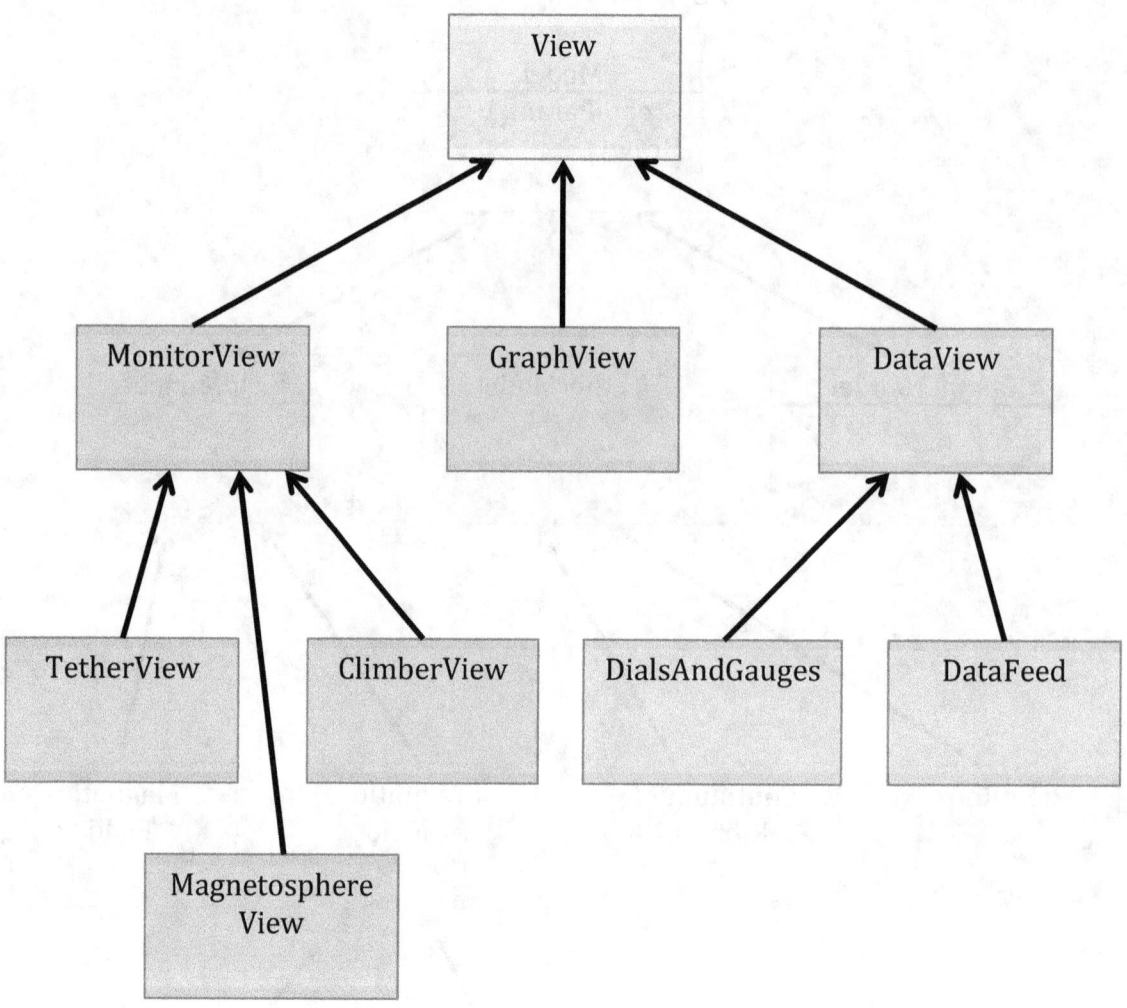

Figure 8. Partial UML diagram of view classes.

Controller Package Controllers allow models and views to be steered by a few parameters exposed to the user. In addition to model and view controllers, there may be run controllers, analysis controllers and test controllers, among others. These are shown in the UML diagram in Figure 9.

There may be several kinds of model controller, such as a console with knobs and sliders for entering model parameters or drop-down menus for choosing models. View controllers will allow users to customize the appearance of model output. This would include changing the scale of the display or graph axes, or reconfiguring the graphical user interface to suit a particular aspect of study.

Run controllers would allow users to set things like start and stop times for the simulation, and perhaps the time increments used by the time steppers when this is not done automatically. Analysis controllers would provide options for post-processing data from the model or plotting it on the screen. Many types of test controller should be made available. These would oversee benchmarking of various models, regression tests and CPU performance profiling.

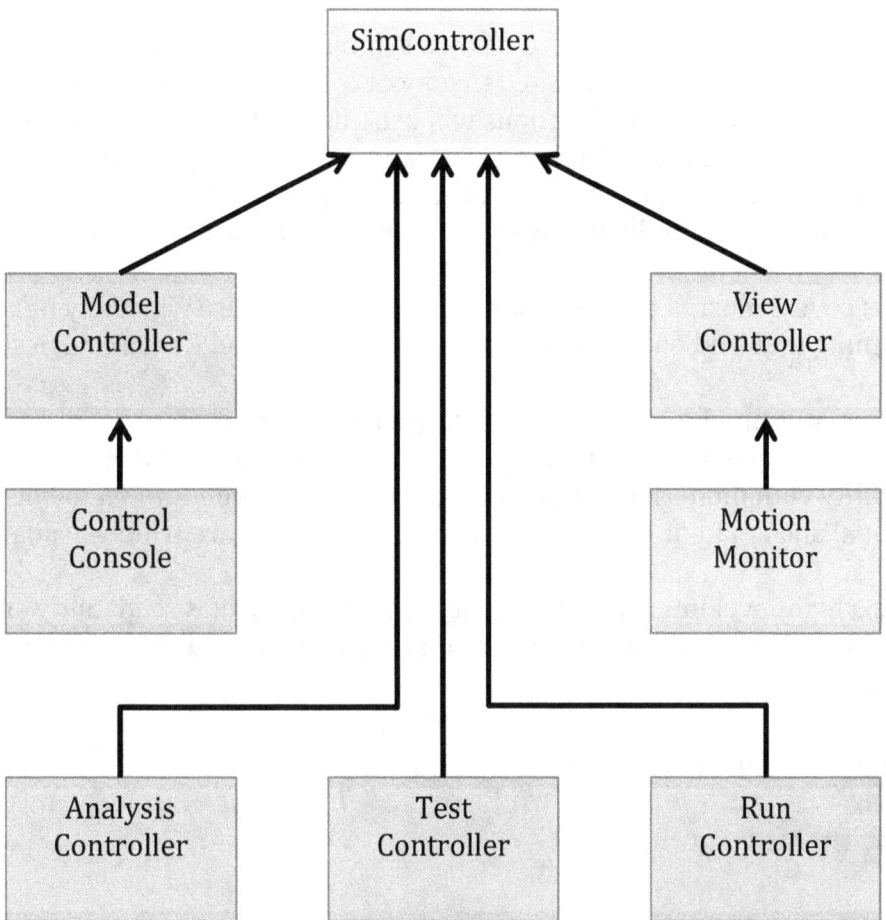

Figure 9. Partial UML diagram of controller classes.

5 Implementation

5.1 Developing the Software

5.1.1 Programming Languages

Before software development can begin, programming and scripting languages need to be chosen. The recommendation of Mathematica and SageMath as the math/physics platforms make certain languages a natural choice.

First of all, Mathematica comes with its own symbolic programming language, the Mathematica Language. It is expected that most of the development of space elevator models and other applications will use this. Mathematica and the Mathematica language are written in C, C++ and Java; its compiled libraries are fully interoperable with external C and C++ libraries. SageMath is written in Python. Its libraries are also compatible with external C and C++ libraries.

While this covers most of the anticipated development, there will likely be programming required outside of Mathematica. Python and C++ are logical choices.

C++ is a versatile, modern programming language which promotes object-oriented design. It has an actively maintained set of standards which ensures its compatibility with modern computing hardware. C++ is well-known to both professional and amateur programmers and its compilers are widely available.

Python is a high-level interpreted language which emphasizes readability and clear programming. It is often used as a scripting language and is popular and well-supported.

Recommendation 6: The programming languages C++ and Python should be available for any simulator development which cannot be done within Mathematica.

5.1.2 Identifying Developers

In order to ensure uniform, high-quality software it is probably best to hire a professional programmer to develop at least the core of the simulator software. This would include the model, view and controller classes mentioned above, interfaces to Mathematica and SageMath software and the infrastructure needed to connect to external databases, among other things. ISEC members should be able to develop new space elevator applications using the core software, and space elevator modelers outside of ISEC should be able to develop the interfaces needed to run and test their applications within the ISEC simulator.

Recommendation 7: The core simulation software should be written by a professional programmer with subsequent development done by ISEC members.

5.2 Using the Software: Concept of Operations

5.2.1 Identifying Users

A high-quality software simulator will attract many users with different needs and qualifications. Users will range from scientists and engineers to students to gamers.

Scientists and engineers will likely want all the functionality that the simulation can offer. This should allow them to carry out detailed research on most aspects of space elevator motion and operation. They will also need to develop and test their own models and applications.

Students will use the software to learn about space elevators and the basic physics behind them. They will use existing models supplied with the simulator. Advanced students may also become developers, producing their own models.

Gamers and casual users might only be interested in using the simulator for fun. Such purposes may include anything from satisfying a curiosity about space elevators to including simulations of them in an online game. Access to the full simulator functionality would likely be unnecessary in these cases.

5.2.2 The User Interface

Users will operate the simulator from their own devices, be it cell-phone, tablet or PC, through one or more interfaces. These can take the form of a command line, file input or graphical user interface (GUI). It is likely that the GUI will be most popular, so some consideration of its look and feel is due.

A notional view of the GUI as it would appear on a user's screen is shown in Figure 10. It is one of many views that would be made available to the user in order to control the simulation. Sliders and knobs would provide analog control over a restricted set of model parameters that the model author has seen fit to provide. More parameters might be made available for scientists and engineers than for casual users. In either case the response of the simulated space elevator to changing parameters should be as fast as possible.

The command line and file input interfaces would be useful for sending simulation results straight to analysis tools without viewing the behavior on a screen or monitor. This mode of operation is well-suited to batch job running.

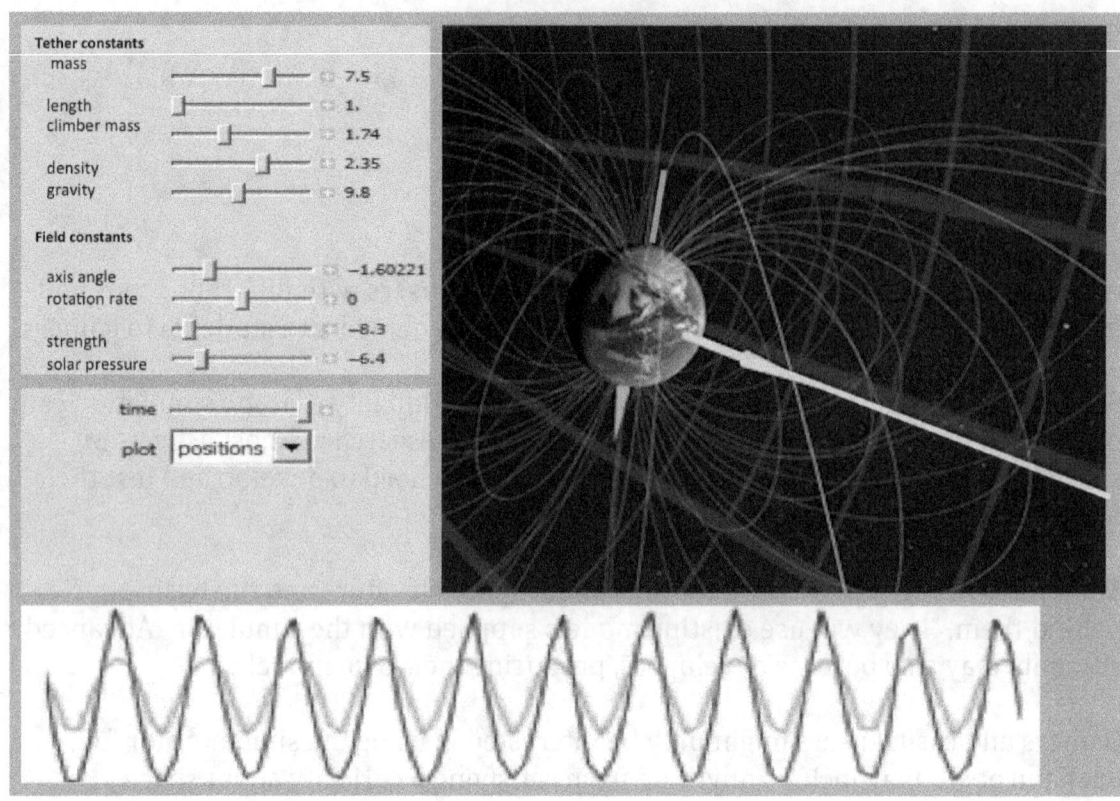

Figure 10. Notional view of user's screen while operating simulator.

5.2.3 Providing and Distributing Resources

Once the simulator software is developed it will become the property of ISEC, which will maintain and distribute it. Software developed for or by ISEC would be under an open-source license such as GNU Public License (GPL). Distribution could be realized in three ways: copies of the source program, pre-built libraries or a ready-to-run instance of the simulator.

1) Copies of the source program could be made available for download. Users would then compile, build and use it on their own computers. Users must first have downloaded their own copies of Mathematica and SageMath, and established access to the necessary physics databases.

2) Pre-built libraries instead could be downloaded, eliminating the need for the user to compile and build the source libraries. As in the first option, users would be responsible for obtaining their own copies of Mathematica and SageMath and database access.

3) With a ready-to-run instance of the simulator, the compilation and building of the software would be done by ISEC, relieving the user of those jobs. No downloading

would be required as an instance would be accessed online through a cloud-based hosting arrangement.

The last option requires that ISEC configure and qualify a cloud platform to support a simulator environment. This would include database access and temporary storage of simulator results. ISEC would also be responsible for obtaining a multi-user Mathematica license in case the open-source option is not chosen. The cost of providing these resources via the cloud has not been studied in detail, but it could be recouped by charging user fees through the use of tokens or other online transactions.

5.2.4 Support for Users and Developers

Simulator users throughout the space elevator community will require access to the simulator software through a key or download facility, access to simulator documentation, help with installing and running the simulator, advice on problem solving and bug fixing, and a place to discuss issues and share results.

A professionally developed web site would fulfill these needs by allowing people to register as simulator users or to subscribe to an online forum where users and experts could post questions and comments. The web site would also serve as a repository for documentation. Web site maintenance and forum moderation could be performed either by volunteers or professionals.

It is also possible that instead of a dedicated web site, the above functions could be accomplished by an existing online service such as GitHub.

5.2.5 Security and Maintenance

Security is a concern with all modern software and especially online applications. The integrity of the simulator software must be assured and the web site must take reasonable precautions against hacking or other attacks. Following good programming practices during the development phase will help with these issues, as will controlling access to the software, vetting users who apply to join the online forum and testing the software contributed by application authors before including it into the simulator.

Maintenance will include the security and upkeep of the web site, its forum and its membership list. It will also include the periodic upgrade of the simulator software so that it remains up-to-date and conformant with advances in computing technology. Simulator software and documentation should be maintained in a version control environment which allows collaborative development and frequent updates. Git is a popular example of such an environment.

5.2.6 Software Testing and Quality Assurance

Delivery of a high-quality space elevator simulator requires that mistakes and flaws be identified and removed before the software is released for general use. This can be accomplished by following best practices during the design phase and by extensive testing of pre-release versions of the software. Quality assurance during testing can take the form of unit testing, system testing, regression testing and benchmarking.

Unit testing checks small segments of software which are responsible for one or a few functions used in the larger simulation. In the space elevator simulator this might be the testing of a function which accesses the value of the Earth's magnetic field from a database and returns it to the main program in a given coordinate system. Unit tests are numerous and typically developed concurrently with the software.

System tests check the performance of large subsets of the software, integrating many sub-systems or packages and subjecting them to tests which require them to work together. One useful system test would be to make sure that the differential equation integrators that predict the motion of the space elevator tether do not produce divergent results after a long period of computation time.

Regression testing is used to ensure that the performance of the software does not decline over time as new features are added or other changes are made. At regular intervals, for example whenever a new version of the simulation software is released, the same system test would be run and its results checked. Any drift of test results away from a standard or optimized set of values would flag a problem.

Benchmarking is critically important for both development and operation of the simulator. The performance of any space elevator application must be assessed against standard data and tests in order to determine whether or not it describes reality with sufficient accuracy. Choosing the most discriminating tests and obtaining real-world data is a challenge, but the more such tests run and the larger and more diverse the dataset compared to, the more robust will be the application.

Recommendation 8: The simulation software should be maintained for the indefinite future, with upgrades, regular testing, security and differing levels of access for a wide variety of users.

6 Roadmap for Development

6.1 Phased Development

A space elevator simulator that meets all the above requirements will be a large project. Given a small to moderate budget, development must proceed in phases, starting with a minimally functioning simulator, then adding sophistication and functionality as time and resources permit. This approach also allows the software to evolve in order to meet changing needs.

Recommendation 9: Develop the simulator software in phases, adding components as time and resources permit.

6.1.1 Development plan

A detailed development plan is beyond the scope of this report, but the following sketch provides the main aspects. A phased plan would begin with the development or procurement of a rudimentary framework to which a few simple space elevator models and a basic testing suite would be added. At this point a preliminary version of the simulator could be deployed for testing and evaluation.

Pending successful testing, a few, more sophisticated models would be added. These may require, for example, detailed gravitational and magnetic field databases. The framework would thus need to be extended so that database access would be provided to the models. A space elevator application developed outside of this simulator should be added to test the ability of the framework to accommodate diverse programs.

Finally, deployment for general user operations should take place. At this point it would be possible for users to run existing applications and analyze results, or develop new applications using the framework.

The major development categories will be the framework, space elevator models and databases. To a large extent these can be developed in parallel, with detail added as needed. Some of the developments required for a mature simulator are discussed below.

6.1.2 Framework

As mentioned above, the framework connects various simulator components and provides access to services. It will provide access to the Mathematica and SageMath tools necessary to build space elevator models, plug-in slots so that natively and externally developed models can be used within the simulator, standard interfaces which allow models and applications to access various physics databases, the environment in which to run space elevator applications and collect output from them, and other utilities such as analysis programs and a scripting language.

6.1.3 Models

The core of the simulator will be a collection of space elevator models of varying complexity. Based on the use cases of section 2.3 and the functional requirements of section 3.1, ten space elevator and environment models have been identified to carry out the bulk of the simulation work:

1. a series of inextensible strings or rigid rods and masses linked by swiveling springs, treated by series of differential equations
2. a continuous, flexible, extensible string with attached masses, treated by differential equations and/or finite element techniques
3. a 3-D representation of the tether volume using continuum mechanics with finite element treatment and non-linear partial differential equation solvers
4. a heat conduction model of the tether volume and material
5. a material model of the tether, including internal structure, possibly down to the molecular level
6. a model of the electric and magnetic fields in the magnetosphere and solar wind
7. models describing radiation transport through matter
8. a model of wind intensity and direction versus altitude and time
9. an aerodynamic model of the tether and tether climbers
10. a gravitational field model including ephemeris data and the effects of Moon, Sun and planets, as well as the effects of a non-spherical Earth

6.1.4 Databases

Many of the models will require access to publicly available databases to provide environmental data such as local gravity, magnetic fields and radiation intensity. Possible candidates are the EGM2008 geopotential database, the Fairfield large magnetosphere database and the AE9 and AP9 radiation field databases. Facilities for handling and accessing the sometimes large number of files in a database will need to be developed.

6.2 Funding the Development

This study does not include a detailed estimate of the cost to develop the simulator software. However, a rough estimate can be made by assuming the work can be done in two years by a software architect, a software developer, a documentation/web page developer/help desk operator and a project manager. See section 6.3 for a proposed division of labor. Accounting for salaries and overhead this comes to about $300,000 per year for two years.

Several options for obtaining these funds have been considered, including crowd-funding campaigns, such as Indiegogo or Kickstarter, funding from private companies, such as Boeing, MicroSoft, or Amazon Web Services, and grants from governments or universities, such as DARPA or National Science Foundation.

The study team opted for the crowd-funding approach as the one most likely to generate funds in the shortest time.

Recommendation 10: Mount a crowd-funding campaign to obtain development funds.

The cost of continued maintenance, user support and security of the developed software was also not considered although it is clear that these expenses will not be negligible. It is possible that they can be funded by user fees, donations and subscriptions.

6.3 Steering Future Development

As mentioned above, a software architect, a software developer and a project manager will be required. The software architect and software developer may or may not be ISEC members, while the project manager should be an ISEC member. During the first three to six months, a software architect will work full-time defining the detailed structure of the project. In the next six to nine months a software developer will work full-time to implement the design, completing 80-90% of the project. In the second year the developer will work half-time completing the remaining 10-20% of the project, ironing out numerous bugs, adapting the software to user requests, and so on. The architect may be kept on call during this period at part-time. Also in the second year a half-time documentation/web page developer/help desk operator will be hired. The project manager will work half-time for the full two years overseeing the development.

After each of the first two years, and in subsequent years, the project should be evaluated and modified as needed. This will be the purview of the simulation committee. Such a committee should be appointed to guide development, allocate resources, request further funding from ISEC and advise the project manager. ISEC board members and the project manager would be logical candidates to sit on this panel. The committee should take input from simulator users, the project manager, the software architect and the software developer.

Recommendation 11: Appoint a simulation steering committee to oversee and guide development.

The simulator will be used for the foreseeable future in space elevator research, development, deployment and operation. Over this time substantial changes to the software will be made as it is used and evolves. It will be up to the steering committee to ensure that the evolution proceeds in a way that maintains quality and versatility.

7 Conclusions and Recommendations

7.1 Conclusions

A software simulator is an essential first step to the development, construction and operation of a space elevator. The simulator should be designed to meet the use cases that arise within these contexts, be developed using modern programming techniques, and use, where possible, existing software.

In addition to the simulation software itself, a collection of other software must be made available to developers and users, including Mathematica and SageMath, on which the simulator will be based, a C++ compiler and a Python interpreter. These must be maintained for the foreseeable future, with upgrades, testing and security.

A wide range of users will access the simulator, from professional developers to users in science and engineering to casual gamers. The needs of this user community will certainly change over time, requiring changes in the simulation software. Flexible, modular and professional software will therefore be required and guidance for its evolution should be provided by an ISEC simulation committee.

Funding the development and maintenance of this project must now be undertaken. Possible means include crowd-funding, grants and user subscription fees, with crowd-funding being the most likely for the time being.

7.2 Recommendations

Based on these conclusions, the study team made 11 recommendations to guide development of the simulator.

1) A software toolkit should be developed which can simulate the space elevator.
2) The simulator should serve and inform the development, construction and operational phases of the space elevator.
3) The design of the simulator software should follow the Model-View-Controller (MVC) concept.
4) A multi-purpose math/physics platform should be chosen upon which the simulation software will be built.
5) Mathematica should be used as the math/physics platform of the simulator, with SageMath as an open-source alternative.
6) The programming languages C++ and Python should be available for any simulator development which cannot be done within Mathematica.
7) The core simulation software should be written by a professional programmer with subsequent development done by ISEC members.

8) The simulation software should be maintained for the indefinite future, with upgrades, regular testing, security and differing levels of access for a wide variety of users.

9) Develop the simulator software in phases, adding components as time and resources permit.

10) Mount a crowd-funding campaign to obtain development funds.

11) Appoint a simulation steering committee to oversee and guide development.

Appendix A   International Space Elevator Consortium

Who We Are

The International Space Elevator Consortium (ISEC) is composed of individuals and organizations from around the world who share a vision of humanity in space.

Our Vision

The ISEC vision is a world with inexpensive, safe, routine, and efficient access to space for the benefit of all mankind.

Our Mission

ISEC promotes the development, construction and operation of a space elevator infrastructure as a revolutionary and efficient way to space.

What We Do

Our main functions include:

- providing technical leadership to promote the development, construction, and operation of space elevator infrastructures
- acting as the "go to" organization for all things space elevator
- energizing and stimulating the public and the space community to support a space elevator for low cost access to space
- stimulating science, technology, engineering, and mathematics (STEM) educational activities while supporting educational gatherings, meetings, workshops, classes, and other similar events to carry out this mission

A Brief History of ISEC

The idea for an organization like ISEC had been discussed for years, but it wasn't until the Space Elevator Conference in Redmond, Washington, in July of 2008, that things became serious. Interest and enthusiasm for a space elevator had reached an all-time peak and, with Space Elevator conferences upcoming in both Europe and Japan, it was felt that this was the time to formalize an international organization. An initial set of directors and officers were elected and they immediately began the difficult task of unifying the disparate efforts of space elevator supporters worldwide.

ISEC's first strategic plan was adopted in January of 2010 and it is now the driving force behind ISEC's efforts. The Strategic Plan calls for adopting a yearly theme to focus ISEC activities. In 2010, ISEC also announced the first annual Artsutanov and Pearson prizes to be awarded for "exceptional papers that advance our understanding of the Space Elevator." Because of our common goals and hopes for the future of mankind off-planet, ISEC became an Affiliate of the National Space Society in August of 2013.

Our Approach

ISEC activities are pushing the concept of space elevators forward. These cross all disciplines and encourage people from around the world to participate. The following activities are taking place in parallel:

- CLIMB – this peer-reviewed journal invites and evaluates papers and presents them in an annual publication with the purpose of explaining technical advances to the public. The first issue of CLIMB was dedicated to Mr. Yuri Artsutanov (a co-inventor of the space elevator concept); and, the second issue was dedicated to Mr. Jerome Pearson (another co-inventor). CLIMB is scheduled for publication each July. Issues can be downloaded at www.isec.org.

- Yearly conference – international space elevator conferences were initiated by Dr. Brad Edwards in the Seattle area in 2002. Follow-on conferences were in Santa Fe (2003), Washington DC (2004), Albuquerque (2005/6 –smaller sessions), and Seattle (2008 to the present). Each of these conferences had many discussions covering the arena of space elevators. Recent conferences have been sponsored by Microsoft, the Seattle Museum of Flight, the Space Elevator Blog, the Leeward Space Foundation and ISEC.

- International cooperation – ISEC supports many activities around the globe to ensure that space elevators progress towards a developmental program. International activities include coordinating with the two other major societies focusing on space elevators: the Japanese Space Elevator Association and EuroSpaceward. In addition, ISEC supports year-long technical studies: research into a single topic to ensure progress in a discipline within the space elevator project. Reports from each yearly study can be downloaded at www.isec.org.

- The first such study was conducted in 2010 to evaluate the threat of space debris. The second study, and resulting report, focused on space elevator operations. The 2013 study focused upon tether climber designs. The 2014 topic is Space Elevator Architectures and Roadmaps. There is one topic chosen for 2015; Earth Port Design Considerations. The products from these studies are reports that are published to document progress in the development of space elevators.

- Symposia are held and presentations made at the International Academy of Astronautics and the International Astronautical Congress each year.

- Competitions – ISEC has a history of actively supporting competitions that push technologies in the area of space elevators. The initial activities were centered on NASA's Centennial Challenges called "Elevator: 2010." Inside this were two specific challenges: Tether Challenge and Beam Power Challenge. The highlight came when Laser Motive won $900,000 in 2009, as they reached one kilometer in altitude racing other teams up a tether suspended from a helicopter. There were also multiple competitions where different strengths of materials were tested going for a NASA prize – with no winners. In addition, ISEC supports the educational efforts of various organizations, such as the LEGO space elevator climb competition at our Seattle conference. Competitions have also been conducted in both Japan and Europe.

- Publications – ISEC publishes a monthly e--Newsletter, its yearly study reports and an annual technical journal [CLIMB] to help spread information about space elevators. In addition, there is a magazine filled with space elevator literature called Via Ad Astra.

- Reference material – ISEC is building a Space Elevator Library, including a reference database of Space Elevator related papers and publications.

- Outreach – People need to be made aware of the idea of a space elevator. Our outreach activity is responsible for providing the blueprint to reach societal, governmental, educational, and media institutions and expose them to the benefits of space elevators. ISEC members are readily available to speak at conferences and other public events in support of the space elevator. In addition to our monthly e--Newsletter, we are also on Facebook, Linked In, and Twitter.

- Legal – The space elevator is going to break new legal ground. Existing space treaties may need to be amended. New treaties may be needed. International cooperation must be sought. Insurability will be a requirement. Legal activities encompass the legal environment of a space elevator -- international maritime, air, and space law. Also, there will be interest within intellectual property, liability, and commerce law. Starting work on the legal foundation well in advance will result in a more rational product.

- History Committee – ISEC supports a small group of volunteers to document the history of space elevators. The committee's purpose is to provide insight into the progress being achieved currently and over the last century.

- Research Committee – ISEC is gathering the insight of researchers from around the world with respect to the future of space elevators. As scientific papers, reports and books are published, the research committee is pulling together this relative progress to assist academia and industry to progress towards an operational space elevator infrastructure. For more, visit http://isec.org/index.php/about-isec/isec-research-committee

ISEC is a traditional not-for-profit 501(c)(3) organization with a board of directors and four officers: President, Vice President, Treasurer and Secretary. In addition, ISEC is closely associated with the conference preparation team and other volunteer members. Address: ISEC, PMB 204, 9272 Jeronimo Rd. Ste. 107A, Irvine, Ca 92618-1978 inbox@isec.org / www.isec.org

Appendix B   ISEC Yearly Study Reports

All past and current ISEC reports are listed here. They are available for sale in hardcopy or at no cost in pdf format at www.isec.org.

2010 – Space Elevator Survivability, Space Debris Mitigation evaluated the threat of space debris and proposed ways to reduce it.

2011 – Carbon Nanotube Developmental Status summarized the progress in carbon nanotube research in meeting space elevator needs.

2012 – Space Elevator Concept of Operations focused on how space elevators would be organized and operated.

2013 – Design Considerations for the Tether Climber studied the physical, operational and economic aspects which constrain space elevator climber design.

2014 – Space Elevator Architecture and Roadmaps delineates the development of the space elevator as a transportation infrastructure.

2015 – Design Considerations for the Earth Port examines the interface of the space elevator and the Earth's surface in terms of a harbor and transportation hub.

2016 - Design Considerations for the Space Elevator GEO Node and Apex Anchor considers the functional aspects of the GEO and Apex Anchor regions in terms of the transportation infrastructure and space enterprise.

2017 - Design Considerations for a Software Space Elevator Simulation studies the requirements for a software space elevator simulator and provides an outline for its development.

Appendix C   Terms and Acronyms

Apex Anchor   The multi-functional complex located at the space end of the space elevator providing counterweight stability as a large mass

C++   A general-purpose, compiled programming language with object-oriented features as well as low-level memory manipulation features

Climber   A payload carrier which ascends and descends the space elevator tether

Cloud computing   An information technology that allows location-independent access to shared, distributed computing resources

Earth Port   The multi-functional complex located at the Earth end of the space elevator providing mechanical and dynamical termination, and port access

Finite element   An approximate method for solving differential equations by dividing complex volumes into small regions which permit exact solutions

GEO   Geosynchronous Earth Orbit

Git   A version control system for tracking changes in, and allowing collaborative development of, software files

ISEC   International Space Elevator Consortium

Java   A compiled programming language that allows the development of applications for use across many computing platforms

Object library   A collection of compiled program units stored in binary format

Object-oriented   Programming methodology centering around spaces in computer memory and promoting modular software development

Open source   Non-proprietary software in which the source program is open to the public

Python   An interpreted programming language often used for scripting and connecting diverse software applications

Tether   A long, thin ribbon of material stretching from the surface of the Earth though GEO to the Apex Anchor, along which the space elevator climber travels

User Interface The means by which a user interacts with a program, a subset of which is the graphical user interface (GUI)

Appendix D   Brainstorming Session Minutes

D.1   Session at 2015 Space Elevator Conference

In order to collect ideas to inform the development of a space elevator software simulator, four subject areas were discussed: customer definition, project scope, simulator model content and capabilities, and software infrastructure. A summary of each discussion area is presented here.

Customers

To define the customers, or users, of the simulator, it is necessary to know who the users would be, how many of them exist worldwide and how the simulator would be used. As the simulator would need to be demonstrated to show its precision and quality, it is likely that the first users would be space elevator investors and the people who are seeking investment. The next users would be the academic researchers and early commercial entities who would perform R&D using the simulator. It may also find a market within the gaming community where space elevators are already in use (KSP).

It is difficult to estimate the number of worldwide users, but it is likely to be small and to constitute a niche market.

The simulator is likely to be used as a software toolkit or as a set of add-on tools to an existing product, like MatLab or Ansys. The toolkit should not be specific to ISEC but be useful to all space elevator researchers, allowing private applications to be built on top of it. The software will have to be licensed in such a way as to allow multiple users and real time analysis for tether operators and developers.

Scope

The development of the simulator should be delineated at the outset, describing what should be modeled and what should not, and what its ultimate applications will be. It would be useful to first survey the availability of existing proprietary and open modeling tools in order to avoid duplication of effort. From this it could be determined what properties the model should have.

The scope of the simulator project can be defined by its products, which currently include models of a pathfinder 1000 km tether experiment, the full tether deployment scheme, its operational mode and several failure modes. The project should begin with a baseline model and evolve to include more detail and complex operations. The parameters of each product should be defined beforehand. Testing and performance monitoring should be included at each major step, with periodic benchmarking against other application software.

Management oversight would be required to ensure that the project stays within scope. For this a Work Breakdown Schedule (WBS) should be developed to map out the work and project timeline. This will be essential for determining project costs.

Model Content

In addition to the physical inputs listed above, the model should include all known external and internal influences on the tether such as motion at the Earth, GEO and apex nodes, gravitational effects of the Sun and Moon, electromagnetic interactions with the magnetosphere, internal and external friction, heating and cooling due to the tether moving in and out of Earth's shadow, solar and atmospheric winds, multiple climbers and thrusters on climbers.

The ability to model various tether deployment scenarios and the deployment of payloads at various points along the tether is essential. The model should in general predict the response to forces at any point along the tether which may be due to oscillations, debris strikes, tides, and so on. It should also be able simulate the application of counter-forces in order to damp oscillations and correct for the effect of impacts or other effects. These calculations should be done sufficiently quickly that they can be used for real-time corrections of tether motion.

As the tether will likely be constructed from multiple materials, the model should be able to simulate the internal stresses due to composite structure.

Any detailed model should include error logging and run-time status reporting.
System control models should contain links to metrology data in order to allow feedback. Simplified models which have analytical solutions should be used to provide sanity checks and a baseline understanding of more complex models. In all cases the mathematical stability of the model must be guaranteed.

It is unlikely that the initial model will contain all the above requirements, so any model developed must be modular and extensible. This will allow new requirements to be implemented as necessary and will help the project remain within scope.

Infrastructure

Software infrastructure must be developed to support the development and operation of the space elevator simulator. It should be 1) multi-platform, 2) multi-developer, 3) multi-user and 4) distributed.

Item 1 requires that all software should run on multiple computer/operating system combinations. Modern computer architectures use multithreading, parallel processing and Graphics Processing Units (GPUs). Because any detailed model will be computing-intensive, its software should take advantage of these features.

Item 2 requires that all space elevator programmers have simultaneous access to development and testing tools, which include compilers or interpreters for programming languages such as C++ and Java, scripting languages such as Python and various profiling tools to optimize and test software. All software and software documentation should reside in a versioning repository, such as GitHub or Perforce Helix Cloud, which enables collaborative software development, provides backup and access to all past versions of software, and release management.

Item 3 allows simultaneous access to several users or operators of the simulator. Easy and efficient use requires a good user interface and professional-grade visualization so that model performance can be quickly monitored and modified. Analysis tools including histogramming and plotting should also be provided. All software should be collected in a toolkit or library of modules which uses a web Application Programming Interface (API) to ease their integration with other applications. The toolkit should be easy to install and provide an embedded user guide. Online help and training would also be useful. All software developed should be open whenever possible.

Item 4 acknowledges that in-house computing farms and storage disks are a thing of the past. The increased availability of cloud computing and data storage from commercial sources is becoming more and more affordable and avoids the necessity of hardware maintenance. Seti@Home is another example of how significant computing power may be obtained at low cost. Because data will be shared, it will be necessary to adopt a standard format for data storage; this could be the simple Comma Separated Value (CSV) format or the sophisticated HDF5 format.

Attributes
For the purpose of project definition, the desired attributes of the simulator applications and the simulator software infrastructure were extracted from the above summaries. The software should be inclusive of all known, non-negligible physics effects, mathematically stable, physically accurate, capable of real-time response, testable and extensible.

D.2 Session at 2016 Space Elevator Conference
Four topics were discussed: what must be simulated, the software tools required to carry out all the simulations, testing the simulations and integrating different software tools into a system.

What must be simulated

Any space elevator model must be flexible and modular. It must provide a common, user-friendly interface which accommodates different models and allows comparative testing of models. Models and interface should have selectable parameters.

Several areas of simulation were identified:

- basic tether behavior, including 3-D modeling of variable linear density, elasticity, longitudinal and transverse oscillations, their damping and the forces they cause in all components of the system
- tether dynamics factors, which include radiation and solar activity, earthquakes, tidal forces from Moon, Sun and planets, Earth nutation, atmospheric and ocean forces, tether material properties and payloads

- non-dynamic factors, such as intermodal transportation, space traffic control and simulation of the swarm of vehicles around GEO, space debris, life support, structural design, communication and environmental and political simulations concerning location of Earth Port and failure/disaster scenarios

Software tools

What is expected of the tools and models must first be clearly defined. When this is done, a survey of available tools should be performed in which apples-to-apples comparisons between tools are made.

Possible tools include:

- modeling tools and physics solvers: Mathematica, MatLab, R, Maple
- general tools: API (application programming interface) or SDK (software development kit), GUI (graphical user interface), IDE (integrated development environment) such as Eclipse, QT Designer, QT Creator or Visual Studio
- compilers/interpreters: strongly typed languages preferred like Java, C++
- wrappers to allow any component to fit into a framework
- software from other domains such as video games, environmental modeling and transportation modeling could be leveraged

Policy and conformity proposals were made. Use of proprietary software should not be shunned but legal aspects of sale and usage must be considered. The open software model should be followed as much as possible consistent with the previous point. There should be freedom to use different tools within common framework. Note that DARPA announced they are requiring use of a common tool. Should a common language or programming style be enforced? An ISEC simulation guru should be appointed.

Finally, a number of unanswered questions were raised. Should all software be submitted to the shared system, and does this raise a conflict with propriety issues? Should source code be demanded for model submissions? Should a top/down or bottom/up modeling approach be followed?

Testing

The simulation must be tested in several ways, including comparison to experiment, comparison to models and quality assurance. There must be a mechanism to identify differences in models that are expected to give similar results and a way to resolve discrepancies between them. Benchmarking, model comparison and quality assurance were identified as major areas of testing.

Benchmarking is the comparison of simulation to physical setups, which could include tethers in mineshafts, large vacuum facilities or vacant amusement parks, tests in wind tunnels, high altitude balloon or rocket tests, Foucault pendula, and on-orbit tests.

Comparison involves evaluating the features and precision of each model and deciding which one best corresponds with reality, including analysis of experiment metrology and resolving differences between models. Evaluation may also include trade studies involving accuracy vs. required CPU time. Quality assurance involves testing the software itself by checking for bugs, unit and integration testing and regression testing. Most of the testing should be automated and may possibly be done using crowd processing such as Galaxyzoo and SETI@Home.

System integration and framework

A software framework is analogous to an electrical bus to which components may be attached and communicate with one another. The many software modules required for simulation will need to be organized, maintained and made to work together by the framework. General purpose frameworks already exist, such as USQUE.
The framework should:

- provide a unified modeling framework, with a consistent architecture; this would mitigate some of the complexity due to different systems interacting;
- provide access to many, possibly large, databases as inputs to models and possibly provide a single data API for these;
- provide a unified GUI;
- allow for modular, object-oriented software so that many different space elevator models can be accommodated;
- support a common format for data input/output to encourage collaborative development;
- be robust and performant: it will be necessary to do long-term, real-time simulations;
- be able to run on the chosen hardware; no problems with existing hardware options were identified.

Project management will be a necessary component of the simulation effort. This will require expertise and collaborative effort. Outreach to interested parties is recommended as is creating a presence for ISEC at software development conferences. It may be useful to hire a software professional to assist with or manage the project.

Agile development methodology should guide the work and keep development focused on achievable goals. Curation of contributed software can follow existing examples in the open-source software community in which there is a software repository with oversight for contributions.

D.3 Session at 2017 Space Elevator Conference
This session addressed four topics concerning the space elevator simulation software: features a user would like to see and how it would be used, features space elevator

application developers would like to see, deployment and management of the space elevator software and funding space elevator software development.

Simulator Features a User Would like to See
We see this as a modular program that starts out with only what we anticipate, but expands over time as users contribute requests for changes/additions to the software.

When opening the software, the user should have the option of language, time zone, and region in accordance with ISO standards (https://www.iso.org). The user should then be able to choose the mode of use, i.e., game play, educator, student, customer, engineer, etc.

An online tutorial would introduce new users to the software and guide them through the available menus. A demo mode would also be provided in which the user could watch a standard simulation to get an idea of what the software is capable of.

After running a unique scenario, the user should have the option of saving the data for himself, or uploading it to a library within the simulation site to share with other end users. Those users could then watch it for new ideas of what the software is capable of. This option should only be for the serious user.

Not every user will have the same experience; what is seen depends on the user. Serious users will have access to more complicated scenarios than the casual user on the site. Paying customers will get more features than a gamer while an engineer on the development team will have access to every feature available in the latest update.

A gamer version, similar to the existing program available at kerbalspaceprogram.com would be fun or useful for education. Simpler versions (Simulation Lite) could be downloaded for use on tablets or even phones. It could be shared in a multi-user mode for teams.

The system should be user-friendly and intuitive. It needs to be compatible with Windows, Macintosh, and Linux systems for a wide range of users. The file format should be fairly universal for uploading and downloading scenarios (perhaps JSON).

A feedback system needs to be included so that the software can be updated, as stated above. Users should be able to submit requests for updates or changes to the program to suit needs that were not previously anticipated. The backlog of requests should be visible so that users may add a "me too" and vote for specific features. Moderators should be able to consolidate similar requests together to prevent duplication with minor variation. Online servers could also provide a reward feedback system where users who upload their scenarios can see how many times their scenario is viewed. The software could also give the option of feedback so that viewers can rate the scenario.

The primary use for the software will be for developing the space elevator to prepare it for deployment and launch. We anticipate it being used by architects, designers, operations managers, accountants, dispatchers/schedulers, maintenance personnel, emergency response crews, safety officers and perhaps more.

A model that could be followed (so as to not re-invent the wheel) is called Systems Tool Kit. Formerly called Satellite Took Kit, it was expanded multiple times to allow engineers and scientists to perform complex analyses of ground, sea, air, and space assets.

Analysis tools should be provided so that the results of a simulation can be understood and displayed in various formats, such as Gantt charts, bar graphs, pie charts, spreadsheets, data heat maps, time series, playback time sliders with various scales (sec, minute, hour, day, etc.) and more : http://guides.library.duke.edu/datavis/vis_types

The software could be similar to a flight simulator in that the simulation could be portrayed in 3-D, virtual reality or mixed reality.

Features a Space Elevator Model Developer Would Like to See
The simulation software should be developed in an iterative manner. The process should be broken down into modules which are developed and tested separately before being combined into the entire simulator. Simulations should be compared to real-world data.

Desired features include identification of owners of software so that they may be contacted for any questions, task coordination software, identification of subject matter experts (software, domain, etc.), using them as a bridge from high level requirements to technical design, and a means to identify results that are not within some bounds of known, actual or expected results.

Many software tests will be required. There will be a need for test data sets and known outputs, small scale (or unit) tests, larger scale (or integration) tests, "domain relevant" testing (different tests at different places) and validation of low-level pieces of software.

Communication with software developers will be needed. Some ways to do this include collaborative use fora (HipChat, e-mail distribution list, or something else), virtual collaboration tools (Slack, GoToMeeting, WebEx) and perhaps mandatory regular meetings for regular communications.

Concept of Operations for Space Elevator Software
The simulation software will have to be run efficiently, maintained, distributed and secured. Issues of workflow and allocation of computing resources need also to be considered. Settling these issues will result in a concept of operations.

One of the first issues to settle is who will use it. Users will be ISEC members, customers, educators and students, and casual observers and gamers.

ISEC members would need more features than say, a gamer. The customer will need to know whether or not the company can afford to use the Space Elevator and can they should be able to identify that they cannot afford NOT to. The simulator should have a cost analysis to compare to an incremental standard rocket launch model. A customer will need certain parameters, but definitely not all of the available ones. Risk factors can be given, but the customer should not have to concern himself with the same parameters as, for instance, a Space Elevator Safety Officer.

To be sure that only serious users have access to the servers that contain the library, the software should verify user identity and the company or organization represented. Abusers of this access can be downgraded by system moderators.

A distribution model for the software would be to make it accessible via the internet with both online and downloadable versions which would depend on usage and computer capabilities.

Documentation in the form of tutorials could be offered: light online versions for casual users and downloadable versions for heavy compute operations.

Workflow issues include building the software, loading it, dealing with and understanding failures, maintenance and security.

Distributed computing must be considered. This included server farms, such as AZURE, and compute farms (perhaps GPUS). seti@home is an example.

Managing and allocating computer resources will be an important task. This could be done using tokens, which would come with a fee for private use. Perhaps some tokens would be free.

Funding the Simulation Software

There is already significant software development occurring for modeling tethers and space elevators. Most of this software is being developed by researchers to support specific projects. Early progress could be made by better coordination of these. A first big win could be simply storing the source code in a central, browse-able repository online.

The first option for developing a monolithic software simulator would be to use grant money from government research programs. A grant writer could be hired to craft the proposal. Usually the fee is a small up-front payment and a percentage of the award. For example, it could be $3,400 and 10%.

We could also go over non-governmental grants. This could include foundations, NGOs, and non-profits. It could also include large corporations, especially aerospace companies.

We could create two projects, a game and a simulator. The game would be funded by investment and produce a profit. The profits from the game would in turn fund the simulator software development. The game could also be used for outreach and education, something that Kerbal Space Program does well for spaceflight in general.

A powerful software development method is the Open Source Software (OSS) model. Large teams of mostly volunteers develop complex software that is used for personal and commercial purposes. Examples include Linux, Firefox, and KiCad. This usually requires a core team that is paid by the sponsoring body, e.g. Mozilla hires a core team to develop Firefox. However, much of the work can be done by volunteers. This could allow us to leverage small resources into a high quality software program. As the software becomes more capable, researchers would be incentivized to contribute to the existing project rather than developing their own, making the software more desirable and creating a virtuous cycle.

Representatives from ISEC could pitch companies and individuals at trade shows, other conferences, or even something like South by Southwest (SXSW) to create this software of their own initiative.

Similarly, we could pitch researchers, who could in turn acquire grant money to develop the software.